Date D

ANIMALS EVERYWHERE

Books by Ingri and Edgar Parin d'Aulaire

ABRAHAM LINCOLN

ANIMALS EVERYWHERE

BENJAMIN FRANKLIN

BUFFALO BILL

COLUMBUS

D'AULAIRES' BOOK OF GREEK MYTHS

D'AULAIRES' TROLLS

DON'T COUNT YOUR CHICKS

GEORGE WASHINGTON

LEIF THE LUCKY

THE MAGIC MEADOW

NILS

OLA

POCAHONTAS

THE STAR SPANGLED BANNER

THE TWO CARS

ANIMALS EVERYWHERE

by Ingri and Edgar Parin d'Aulaire

DOUBLEDAY & COMPANY, INC., GARDEN CITY, NEW YORK

ISBN: 0-385-07216-3 TRADE
0-385-07703-3 PREBOUND

Library of Congress Catalog Card Number 54-5904

 Lithographed in the U. S. A.
9 8 7 6 5

ANIMALS EVERYWHERE

Far to the South, where it is very hot, live the animals that like very hot weathe
There the lively Monkey hides in the leaves. The shy Zebra gallops over the gras
The fat Hippopotamus rises yawning out of the river. The haughty Giraffe pee
over the tallest trees. The Ostrich fluffs its fine tail feathers.

The swift little Antelope runs to save its life. The Chimpanzee chatters in the treetops. The Anteater uses its long nose to dig up anthills. The wise Elephant is so strong he can uproot trees with his trunk. The little Mouse darts hither and thither.

The Mouse squeaks. The Monkey chatters. The Antelope cries.

The Anteater sniffles. The Elephant trumpets.

The Giraffe is mute. The Zebra squeals.

The Ostrich booms. The Hippopotamus bellows.

The ferocious Tiger jumps to attack its prey. The Dangerous Lion is the king of the beasts. The Ape sometimes walks upright and is much like a man. The Rhinoceros defends himself with the horn on his nose.

The Dromedary with one hump, the Camel with two, can walk for weeks without drinking. The Crocodile lies still like a log. Where these animals live it is never cold and never winter. Sometimes the rain comes down in torrents, but most of the time the burning sun scorches.

The Camel and the Dromedary groan. The Parrot talks.

The Crocodile smacks its jaw.

The Rhinoceros snorts. The Tiger snarls.

The Ape screeches. The Lion roars.

The animals that like weather neither too hot nor too cold live where we live. Here the shaggy Buffalo roams. The Deer plays. The sly Fox sneaks. The Skunk defends himself with his smell.

The noble Horse trots. The proud Cock stalks. The busy Hen scratches. The lively Squirrel gathers nuts. The Donkey balks. The Cat prowls.

The Donkey brays. The Horse whinnies. The Squirrel scolds.

The Cat miaows. The Cock crows.

The Fox yaps. The Skunk rattles.

The Deer bugles. The Buffalo lows.

The Goose waddles. The friendly Cow chews its cud. The mischievous Billy Goat bucks. The Eagle soars through the air. The peaceful Sheep grazes. The Duck walks on the ground, swims in the water, and flies through the air. The Dog hunts.

The sleepy Bear hibernates in winter. The busy Beaver builds dams. The Moose lives in the deep forests. Where these animals live it is cold in winter and warm in summer.

The Moose coughs. The Bear grumbles.

The Beaver splashes. The Eagle screams.

The Dog barks. The Sheep baas. The Cow moos.

The Duck quacks. The Goat bleats. The Goose honks

Far to the North live the animals that like cold weather. The Reindeer digs for grass and moss under the snow. The cruel Wolf pursues his prey. The timid Hare tries to hide. The black Raven steals (like a raven).

The Owl is very wise. The horned Narwhale swims through the icy waters. The sleek Seal crawls up on ice floes to rest. The Eider Duck gives us eiderdowns.

The Eider Duck squawks. The Owl hoots.

The Seal barks. The Narwhale swishes.

The Raven croaks. The Wolf howls.

The Reindeer pants. The Hare thumps.

The Walrus has a very thick hide. The Sea Lion is heavy and clumsy. The Whale spurts water out of a hole in his head.

The Sea Gull sails through the air and dives into the sea for fish. The Polar Bear is a ferocious beast. Where these animals live there is snow and ice all through the year. In summer the sun never goes down and in winter it never rises.

The Polar Bear growls. The Gull cries.

The Whale blows, and he is the biggest

The Sea Elephant roars.

of all the animals on land and in the sea. The Walrus snores.

After the publication of OLA in 1932, the work of Ingri and Edgar Parin d'Aulaire has needed no introduction—their beautiful picture books have delighted countless children ever since.

Ingri Mortenson and Edgar Parin d'Aulaire met in Munich where both were studying art. Ingri had grown up in Norway; Edgar, the son of a noted Italian portrait painter, was born in Switzerland and had lived in Paris and Florence. Shortly after their marriage they moved to this country and began to create the picture books that have established their reputation for unique craftsmanship. Ingri and Edgar work as a team on both art and text. In their early books, effects of color and texture were achieved by the painstaking traditional method of drawing directly on lithographic stone; the same fine results have been achieved in their most recent books by adapting this process to acetate separations.

The d'Aulaires live on a farm in Wilton, Connecticut, with their teen-age son Martin. Their other son and his wife live near by.